# 中国林业碳汇项目审定和核查指南

武曙红　宋维明　主编

中国林业出版社

## 图书在版编目(CIP)数据

中国林业碳汇项目审定和核查指南 / 武曙红,宋维明主编. —北京:中国林业出版社,2012.11
 ISBN 978-7-5038-6868-9

Ⅰ.中… Ⅱ.①武… ②宋… Ⅲ.①森林-二氧化碳-资源利用-项目-中国-指南 Ⅳ.①S718.5-62

中国版本图书馆 CIP 数据核字(2012)第 298632 号

E-mail:forestbook@163.com 电话:(010)83224477
网址:lycb.forestry.gov.cn
发行:中国林业出版社
印刷:北京北林印刷厂
版次:2012 年 11 月第 1 版
印次:2012 年 11 月第 1 次
开本:787mm×960mm 1/16
印张:5
字数:66 千字
印数:1~2 000 册
定价:35.00 元

# Validation and Verification Manual

## for forestry-based carbon sequestration projects in China

**Wu Shuhong**
**Song Weiming**

北京市支持中央在京高校共建项目资助
(项目号：200-670030)
中央高校基本科研业务费专项资金资助
(项目号：200-1240711)
国家重点基础研究发展计划("973"计划)资助
(项目号：2010CB955107)

# 前　言

随着全球气温的不断上升和世界范围极端气候事件的频繁发生，由人类活动产生的温室气体排放引起的全球变暖问题，已引起国际社会的极大关注。"采取行动减少二氧化碳排放"已成为国际气候谈判议程中的一个重要的议题。从1992年联合国环境与发展大会的"边会"（side event）上首次提出"全球碳排放贸易方案"，到《京都议定书》生效后的今天，"碳贸易"的概念已从当时仅为部分科学家和环保主义者所关注的应对气候变化的金融解决方案，发展到作为实现可持续发展、减缓和适应气候变化、灾害管理三重目标的低成本途径和核心经济手段被各国政府纳入国家的低碳发展战略。

自《联合国气候变化框架公约》第15次缔约方会议前夕，中国政府对外宣布，到2020年中国单位GDP $CO_2$ 排放比2005年下降40%~45%的自主减排目标以来，我国政府不遗余力地推动节能减排与低碳转型的实践：中共中央十七届五中全会通过的《关于制定国民经济和社会发展第十二个五年规划的建议》中，明确提出要积极应对气候变化，把大幅度降低二氧化碳排放强度作为约束性指标；在国务院下发的《关于加快培育和发展战略性新兴产业的决定》中也首次将"碳交易"纳入官方文件；在国家发展和改革委员会发布的《关于开展低碳省区与低碳城市的试点工作的通知》中，明确提出要在全国"五省八市"开展低碳省区与低碳城市试点工作。这些官方文件的制定均表明了中国政府在应对气候变化问题上的信心和决心。虽然中央政府目前还没有明确地将具体的碳减排指标分解到地方，但国家在"十二五"规划中对碳强度及排放指标的约束,决定了"十二五"期间各地方政府和各个行业均有承诺减少碳排放减排指标的可能性。目前,中国各相关管理部门已经提出了未来五年将在中国特定的区域和特定的行业开展碳交易的计划。为了顺应国家发展低碳经济的趋势,各地方政府根据各自的地方利益、综合实力、发展水平以及自然资源状况,积极探索利用市场机制促进节能减排的政策和措施。鉴于在《京都议定书》清洁发展机制的碳交易市场中获得的启示和经验,连接现代金融业和低碳产业的碳金融行业,已经受到了我国政府和企业的极大关注。碳金融已被作为应对气候变化的主要激励机制和解决方案,成为各地方政府推动低碳发展重要的路径选择。

我国作为全球碳排放量最大的国家,在利用市场机制促进节能减排的行动中,北京、上海等地利用自身的区位优势,先后成立了北京环境交易所、北京林权交易所、上海环境能源交易所以及天津排放权交易所等碳金融服务机构。随着国家"十二五"规划相关政策的实施,发展林业碳汇市场也将被作为一种低成本的应对气候变化行动的激励机制和解决方案,成为实现2020年我国森林面积增加4000公顷,森林蓄积量增加13亿立方米战略目标的重要路径选择。

我国是第一个成功注册清洁发展机制(CDM)碳汇项目的国家,但这只意味着我国在CDM碳汇项目方法学以及项目设计方面技术的成功,而无论是京都市场还是非京都市场的碳汇交易都是市场经济条件下的市场行为,我国在未来国际碳汇市场中所占的份额不仅取决于增汇技术和潜力,更多的还取决于碳汇项目本身的质量、所采用的审定、核查、认证标准以及公众对审定和核查机构权威性的认可度。虽然《联合国气候变化框架公约》缔约方会议、国际碳排放联盟以及生物多样性保护联盟等机构和组织相继开发了CDM林业碳汇项目方式和程序、AFOLU – VCS、CCBS、

CFS 以及 CCX 等与林业碳汇项目开发、审定、核查以及核证的标准，但这些标准对于缺乏激励机制和市场机构，仍处于起步阶段的我国林业碳汇市场而言，仍不具备应用条件和能力准备。

虽然自 2008 年以来，在中国绿色碳基金、中国绿色碳汇基金会、中国石油股份有限公司以及北京林权交易所等碳金融服务机构的支持和推动下，北京、内蒙古、河北等地已逐步开始实施自愿林业碳汇项目，在中国开启了自愿碳减排市场的新篇章。目前，中国绿色碳汇基金会也正在积极推进这批项目产生的林业碳信用的交易，但由于目前国家碳汇管理办公室对此类市场还没有制定出任何规范的交易体系和核查/认证林业碳汇项目的标准，对审定、核查、核证国内林业碳汇项目的机构的授权或指定也没有严格的审批程序和资质要求，成功交易这些林业碳信用仍还面临着巨大的挑战。

随着国内林业碳汇市场规模的进一步发展，审定和核查标准的缺乏不仅会导致国内林业碳汇市场交易中不同林业碳汇项目之间的林业碳信用缺乏可比性，使林业碳汇市场秩序处于混乱状态，还将影响我国林业碳信用的信誉度和国内外林业碳信用投资者和购买者对林业碳信用的信心，使我国的林业碳汇项目在国际林业碳汇市场中失去竞争力。因此，在顺应国际林业发展的潮流、实现我国地区资源优化配置的背景要求下，北京林业大学碳汇计量与监测中心精心策划并组织国内林业碳汇领域的专家参与编撰了本书，本书集合了我国林业碳汇领域业内权威机构和资深专家的观点，相信对从事与林业碳汇相关业务的政府机构、企业、科研院校、核查机构、咨询机构等人员具有重要的参考价值。

编　者<br>2012 年 4 月于北京

# Preface

As the increase of the global temperature and the frequency of extreme climate events, the problems of global warming coursed by human activities, in form of emission of carbon dioxide has aroused wide attentions. "Take actions to reduce the emission of carbon dioxide" has caused international society' great attention. From first global carbon emission trading strategy that was proposed on UNCED side event in 1992 to Kyoto Protocol which was in effect in 2005, "carbon trading" has already been different from just a financial solution of dealing with climate change which focused by some scientists and environmentalists, and now becomes a low-cost way and core economics means of achieving sustainable development, mitigation climate change, emergency of management, and this approach has been used by many countries as a low-carbon development strategy.

From the eve of the 15th session of the conference of Parties, Chinese government announced that, the goal of China' emission reduction is a 40% -45% reduction in carbon intensity ($CO_2$ emissions per unit of GDP) by 2020 compared with 2005, our government spare no efforts to promote energy saving, emission reduction, and practice Low Carbon Transformation : The Twelfth Five-Year Plan for Economic and Social Development, passed by the fifth plenary session of the 17th CPC Central Committee , proposed that we must actively respond to climate change and make carbon dioxide emissions intensity reduction as an obligatory target; On Speeding up the Cultivating and Developing Strategic New Industry Decision which was issued by the State Council, points out that "carbon trading" need to be included into the official documents; The Notice about Starting Low Carbon Provinces and Low Carbon Cities Pilot Projects, issued by China's Reform and Development Commission, proposes that we need carry out low carbon provinces and low carbon cities of pilot projects in "five provinces and eight cities". These documents indicate that Chinese government' confidence and determination on dealing with climate change. Although central government hasn't allocated the emission reduction targets yet, the index in "the twelfth five-year plan" determines the possibility of reducing emission in every province and industry. Now the relative management section proposed the plan of reducing emission in China' spatial areas and industries in the next five years. In order to comply with the trend of the low-economics development, based on the local interest, comprehensive strength, development level, local governments are starting to explore the strategy of energy saving and emission reduction. In the consideration of the revelation and experiences in carbon trade market of CDM, carbon financial industry, which combines with modern financial industry and low-carbon industry, has became a great choice of promoting low-carbon development of every local government.

As China is the largest contributor of carbon emission, in the action of using market mechanism as the solution of energy saving and emission reduction, China's government taking location advantage of Beijing and Shanghai, successively established Beijing environmental exchange, Beijing forest exchange, Shanghai environment energy exchange and Tianjin climate exchange. With the practice of the twelfth five-year plan, the development of forestry carbon market will be viewed as an incentive mechanism and solution strategy of dealing with climate change in a low-cost way. And this would become a great choice of achieving expanding forest coverage by 40 million hectares and forest volume by 1.3 billion cubic meters.

China is the very first country witch successfully registered CDM forestry carbon project, but this only means that we are successful in developing methodology and project design technology of CDM forestry carbon projects. Either in the Kyoto Carbon Market or Voluntary Carbon Market, the forestry carbon trade is market behavior under the market economy conditions. The share of the market for China in the future international carbon market is not only determined by carbon sink increasing technology and potential, but also by the quality of carbon sink projects, the certification standards, verification and certification, the recognition of verification institutions authority by the public. Though the United Nations framework convention on climate change conference of the Parties, international carbon emissions alliance, biological diversity protection union has developed the methods and procedures of CDM forestry carbon projects, such as AFOLU-VCS, CCBS and CFS, these standards are lack of incentive mechanism and market institutions. And for our elementary forestry carbon market, these standards have no application conditions and capability.

Although since Beijing, Inner Mongolia and Hebei began practicing voluntary forestry carbon projects which are backed by carbon financial service institutions such as CGCF, Chinese Petroleum Corporation, China has opened a new chapter of voluntary forestry carbon market. Now CGCF is dedicated to promote the trades of forestry carbon credit of these projects, but there are no standards to restrain cap-and-trade system, verification and certification standards, plus there are no procedures for examination and approval when we need to improve an institution to examine and approve, so these forestry carbon credits are still facing a fudge challenge.

As the development of China' forestry carbon market, the lack of standards of verification and certification will result in the lack of comparability between different forestry carbon projects in China, and disordering the forestry carbon market sequence, and this would impact the creditworthiness of China' forestry carbon credits and the confidence of the domestic and overseas investors and buyers, and finally make China' forestry carbon projects lose their competitiveness in international forestry carbon market. So under the request of complying with the tide of international forestry development and achieve China' local resources optimization allocation, Carbon Measurement and Monitoring Center of Beijing Forestry University organized the authorities in China' forestry carbon field to compile this book. This book gathered the opinions of authoritative institutions and senior experts, and we believe this book has great reference value for many people who work for the institutions related to forestry carbon, such as government institutions, companies, research institutions, verification institutions and advisory bodies.

Editor<br>
Beijing<br>
April, 2012

**缩略词**

AFOLU：农业、林业和其他土地利用

CDM：清洁发展机制

CCBS：气候、社区和生物多样性标准

CFS：林业碳标准

CCX：芝加哥气候交易所

GHG：温室气体

IPCC：政府间气候变化专门委员会

VCS：自愿碳标准

**Acronyms**

AFOLU：Agriculture, forestry and other land uses

CDM：Clean Development Mechanism

CCBS：Climate, Community and Biodiversity Standard

CFS：Carbon Forestry Standard

GHG：Greenhouse Gas

IPCC：Intergovernmental Panel on Climate Change

VCS：Voluntary Carbon Standard

# 目　录

前　言
1　范围和规范性引用文件 ·················································· 1
　1.1　范　围 ································································· 1
　1.2　规范性引用文件 ······················································ 1
2　术语和定义 ······························································· 2
3　审定/核查原则 ···························································· 4
4　审　定 ····································································· 5
　4.1　审定目的 ······························································ 5
　4.2　审定途径 ······························································ 5
　4.3　审定方法 ······························································ 5
　　4.3.1　审定内容 ······················································· 6
　　4.3.2　审定的步骤 ···················································· 7
　4.4　审定意见 ····························································· 13
　4.5　审定报告 ····························································· 14
　　4.5.1　审定内容的报告要求 ······································· 14
　　4.5.2　审定报告 ····················································· 17
5　核　查 ···································································· 18
　5.1　核查的目的 ·························································· 18
　5.2　核查的途径 ·························································· 18
　5.3　核查的方法 ·························································· 19
　　5.3.1　核查内容 ····················································· 20
　　5.3.2　核查的步骤 ·················································· 20
　5.4　核查报告 ····························································· 22
　　5.4.1　审定报告各项内容及要求 ································· 22
　　5.4.2　核查报告 ····················································· 23

# CONTENTS

1 Scope of Application and Normative References ·········· 25
   1.1 Scope of application ·········· 25
   1.2 Normative references ·········· 25
2 Terms and definitions ·········· 26
3 Principles for validation and verification ·········· 29
4 Validation ·········· 31
   4.1 Objective of validation ·········· 31
   4.2 Validation approach ·········· 31
   4.3 Means of validation ·········· 31
      4.3.1 Lists of validation ·········· 33
      4.3.2 Validation step ·········· 33
   4.4 Validation Opinion ·········· 42
   4.5 Validation Report ·········· 43
      4.5.1 Report requirements for validation list ·········· 43
      4.5.2 Validation report ·········· 48
5 Verification ·········· 48
   5.1 Objective of verification ·········· 48
   5.2 Verification approach ·········· 48
   5.3 Means of verification ·········· 49
      5.3.1 Verification list ·········· 50
      5.3.2 Verification of compliance ·········· 51
   5.4 Verification report ·········· 54
      5.4.1 Report requirements for verification list ·········· 54
      5.4.2 Verification report ·········· 56
References ·········· 57

# 1 范围和规范性引用文件

## 1.1 范　围

本指南适用于在中国实施的林业碳汇项目的审定和核查。

## 1.2 规范性引用文件

下列文件所包含的条款通过本指南的引用而构成为本指南的条款。凡是标注日期的引用文件，其随后所有的修改单（不包括勘误的内容）或修订版均不适用于本指南。凡是不标注日期的引用文件，其最新版本适用于本指南。

——《IPCC 土地利用、土地利用变化和林业优良做法指南》；

——《IPCC 2000 优良做法指南和不确定性管理》；

——《IPCC 2006 国家温室气体清单指南》；

——《CDM 审定和核查手册》；

——《CDM 造林再造林项目方式和程序》。

# 2 术语和定义

下列术语和定义适用于本指南。

**林业碳汇项目(forest-based carbon sequestration project)：**

以增加和维持森林碳储量为目的的造林项目再造林项目、可持续的森林管理项目和避免毁林项目。

**森林(forest)：**

面积大于 0.05～1.0hm$^2$、林木冠层覆盖度(或立木度)10%～30%、就地生长成熟时最低树高可达 2～5m 的土地。冠层覆盖度尚未达到 10%～30% 或树高尚未达到 2～5m 的天然和人工幼龄林也属森林。

**造林(forestation)：**

通过栽植、播种或人工促进天然下种方式，将不曾为森林的土地转化为有林地的直接人为活动。

**再造林(reforestation)：**

通过栽植、播种或人工促进天然下种方式，将曾经为林地的非林地转为有林地的直接人为活动。

**植被恢复(revegetation)：**

通过建立最小面积为 0.5hm$^2$ 的植被，但不满足造林和再造林定义的增加立地碳储量的直接人为活动。

**可持续森林管理(sustainable forest management)：**

能够维持森林健康及生物多样性与社会对林产品日益增长的需求的森林管理活动。

**避免毁林：**

将林地转化为非林地的人为活动。

**项目边界(project boundary)：**

项目参与方控制范围内的林业碳汇项目活动的地理范围。

**碳库（carbon pool）：**

碳库包括地上生物量、地下生物量、凋落物、粗木质残体和土壤有机碳。

**基线情景(basline)：**

能合理地代表没有林业碳汇项目活动时所选择的计量碳库中可能出现的碳排放或碳吸收水平的情景。

**实际净温室气体汇清除(Actual net greenhouse gas removals by sinks)：**

实际净温室气体汇清除是指项目边界内碳库中可核查的碳贮量变化之和，减去项目边界内由CDM造林或再造林项目活动引起的、以$CO_2$当量计算的温室气体源排放的增加。

**泄漏(Leakage)：**

泄漏是指发生于CDM造林或再造林项目活动边界之外的、由造林或再造林项目活动引起的、可测定的温室气体源排放的增加。

**人为净温室气体汇清除(Net anthropogenic greenhouse gas removals by sinks)：**

人为净温室气体汇清除是指实际净温室气体汇清除减去基准净温室气体汇清除，再减去泄漏。

# 3 审定/核查原则

审定和核查机构在进行审定和核查及准备相应的报告时应当遵循以下原则：

（1）一致性。为公平对待所有项目，审定和核查机构应该采用统一的标准对待具有相似特征（例如计量和监测方法学、技术、项目活动时间或区域等）的项目。在整个过程中或者不同项目之间，对待专家的意见采用统一标准。如果政府发布新的决定或标准，审定和核查机构应该主动考虑和制定如何继续与政府的决定保持一致性的可行对策。

（2）透明性。为了避免审定和核查过程中的暗箱操作问题，审定和核查机构所出具的审定和核查报告涉及的信息应该基于文件证据以公开、清晰、客观、中立和一致的方式进行阐述。在遵循该原则时，应该要求项目参与方提供有相应文件支撑的证据说明其透明性。

（3）独立性。审定和核查机构必须对所碳汇项目保持独立性，使审定和核查活动不存在任何偏见以及利益冲突，确保审定和核查结论是基于客观证据得出的。如果审定和核查机构与所审定和核查的项目有着直接或间接的利害关系，公众将有理由质疑其审定和核查结果的公正性。

（4）公平性。审定或核查机构应当准确、真实地反映审定或核查活动、结果、结论以及报告。对审定或核查过程中面临的障碍以及尚未解决的审定或核查者、管理部门以及客户之间的意见分歧进行报告。

（5）保密性。审定或核查机构应当确保审定或核查过程中获得的所有信息的机密性。

# 4 审 定

## 4.1 审定目的

根据国家林业局碳汇管理办公室制定的相关规定或其他碳汇项目实施标准(CDM、VCS等)的要求，对被提交作为碳汇项目注册的拟议项目做彻底、独立的评估，以确保任何提交注册的碳汇项目都符合国家林业局碳汇项目管理部门对注册碳汇项目的要求。

## 4.2 审定途径

审定机构在执行审定工作的过程中，应当：

(1)利用除项目业主提供的证据外的其他所有可获得的所有客观证据，评估项目设计文件中所声明的完整性、准确性和保守性。

(2)确保不能遗漏可能改变审定意见的证据，并确定项目活动是否符合国家林业局碳汇项目管理办公室(或CDM/VCS)制定的相关规定的要求以及该部门批准的方法学和指南的使用条件。

## 4.3 审定方法

(1)审定机构应审定项目参与方提供的所有信息。

(2)审定机构应采用相关国际标准或相关管理部门制定的审查技术来审定项目业主提供的信息的准确性，主要包括：

(a)文件评审。

(i)审定项目参与方所提供的数据和信息的准确性、可信性。

(ii)将所提交文件中提供的信息和数据与下列信息交差进行复核：

——国家林业温室气体清单；

——国家森林统计资料；

——生长模型和材积表；

——航空照片、卫星影像和地图；

——土地利用或土地覆被的历史资料；

——分层和抽样的方法；

——IPCC 好的做法和指南；

——国家土地法和森林法以及项目所在地的相关法律法规。

(b)跟踪行动(例如现场访问，电话或电子邮件访谈)包括两项内容。

(i)访问项目所在地利益相关群体以及了解项目设计和实施的相关人员。

(ii)交叉检查被访者提供的信息，以确保审定过程中没有遗漏相关的信息。

(3)参考与被审定项目活动类似的项目或技术的参考资料。

(4)评审所用方法学中公式的适宜性以及计算的准确性。

## 4.3.1 审定内容

(1)项目设计文件，包括：

——拟议林业碳汇项目的边界；

——计量碳库的选择；

——拟议碳汇项目的土地的合格性；

——拟议项目解决温室气体汇清除非持久性的方法；

——拟议项目活动选择的方法学，包括基线的识别、估计基准和净人为温室气体汇清除数量时使用的计算方法和公式；

——确定管理活动的时间，包括收获循环和核查；

——监测计划；

——当地利益相关群体的咨询过程；

——社会经济学和环境影响，包括对生物多样性和自然生态系统的影响。

（2）拟议项目可行性研究报告。

（3）拟议项目环境影响评估报告。

（4）拟议项目实施方案。

### 4.3.2 审定的步骤

#### 4.3.2.1 项目描述

（1）审定要求。审定机构应当确认项目设计文件是否对拟议的林业碳汇项目活动进行了描述，并能让读者清楚地了解到项目的性质及其相关技术方面的内容。

（2）审定方法。审定机构应当确认项目设计文件是否对下列内容进行了描述：

- 项目活动描述；
- 项目参与方的具体信息；
- 项目活动所在位置；
- 项目活动所在区域的地形、地貌、生态系统以及社会经济状况；
- 拟议林业碳汇项目活动所采用的技术；
- 土地权属和碳汇权属；
- 土地合格性；
- 描述非持久性的方法；

- 碳汇活动碳汇的计入期；
- 估算基线情景和项目情景碳储量的方法；
- 监测计划；
- 碳汇项目活动的社会经济和环境影响；
- 利益相关方反馈的意见。

#### 4.3.2.2 项目边界

(1) 审定要求。审定机构必须确定参与方提交给碳汇项目主管部门的项目设计文件是否对项目业主拟实施或控制的林业碳汇项目活动所在的所有地块的边界进行了描述。

(2) 审定方法。

(a) 通过文件评审和/或访谈确定项目设计文件中是否描述了包括项目所涉村镇的坐标以及项目各地块边界的拐点坐标的项目边界。

(b) 通过文件评审和/或访谈确定项目参与方是否已经：

- 制定了管理制度对拟议项目活动涉及的所有区域内的与碳汇相关的活动（项目参与方控制下的活动）进行管理或控制；
- 对项目参与方控制下的活动进行了管理。

(c) 审定机构应该确认项目参与方控制了拟议项目活动边界内的与碳汇相关的活动。这种控制至少应当包括利用在我国法律体系下可以接受的方式定义排他权，以进行增加或维持森林碳汇活动，从而获得碳汇的人为净温室气体汇清除量。如果要审定的文件和访谈的人员或实体的总数量不少于10，那么审定机构可以申请采用抽样的方法。

#### 4.3.2.3 碳库的选择

(1) 审定要求。审定机构应该确认项目参与方选择的碳库与所选方法学中的相关规定相一致。

（2）审定方法。审定机构必须确定项目设计文件中所选择的拟议项目活动要考虑的碳库是否符合所选择的经批准的方法学的要求。如果所选择的经批准的方法学中允许忽略某些碳库，审定机构必须确定项目参与方是否提供了可以核实的信息来证明此选择，审定机构还必须确保项目设计文件中所提及的所有文件都被正确地引用和解释。如果相关，审定机构必须复核项目设计文件中提供的信息与公共资源或当地专家获得的其他信息相一致。

#### 4.3.2.4　土地合格性

（1）审定要求。审定机构应该确认拟议的项目活动边界的土地是否满足我国林业碳汇项目主管部门（国家林业局碳汇项目管理办公室）对实施林业碳汇项目的土地合格性要求。

（2）审定方法。审定机构应该通过对项目参与方所提供的信息以及现场评审，确认实施拟议项目活动的土地满足我国林业碳汇项目主管部门对林业碳汇项目土地合格性的要求。

#### 4.3.2.5　减少非持久性和泄漏的风险

（1）审定要求。审定机构应当确认项目参与方制定了减少项目碳信用非持久性和泄漏风险的方法或措施。

（2）审定方法。审定机构应当通过评审项目设计文件来确保项目参与方减少非持久性和泄漏风险的方法或措施符合国家林业碳汇项目管理部门或 IPCC 好的做法和指南的相关规定。审定机构应当审定项目设计文件是否描述了"如何解决由于采伐、火灾、病虫害、毁林等人为或自然的原因导致项目产生的碳汇发生碳逆转的问题"。例如，是否将部分碳留存在所产生的碳汇量中暂时扣除等方法。

### 4.3.2.6 项目活动中所采用的方法学

(1) 审定要求。

(a) 审定机构应当确定项目参与方选择的方法学是国家碳汇项目管理部门批准/或建议的版本，或是《IPCC 土地利用、土地利用变化和林业好的做法和指南》建议的方法学。

(b) 审定机构应该确定项目参与方描述的基线情景能合理地代表没有拟议项目活动时，项目边界内温室气体的汇清除量。

(c) 审定机构应该参照所选定的方法学确认项目设计文件中计算人为净温室气体汇清除、基准温室气体汇清除以及泄漏的步骤、方程以及参数符合所选方法学的要求。

(2) 审定方法。

(a) 参照国家林业局碳汇管理部门批准的或者《IPCC 土地利用、土地利用变化和林业好的做法和指南》建议的相关方法学，确定项目参与方正确地选用了相应的方法学。

(b) 审定机构还需根据项目所在地的情况和林业专业知识将项目参与方所提供的选择方法学的证据和其他信息对比，确认这些证据和当地以及林业领域内的常数、数据、信息吻合，否则需要参与方对不同之处进行解释。

(c) 审定机构在审定所选定的方法学是否适用时，需要确定项目是否会引起所选定的方法学规定之外的温室气体排放。

(d) 审定机构应该审定项目设计文件是否列出了方法学中给出的所有可能的基线替代方案，是否对排除每一个不合理的替代方案的理由以及没有拟议的碳汇造林项目是可能会发生的活动进行了论述。如果有，审定机构需要根据自己在专业和地区的知识判断项目设计文件中的论述是否合理。审定机构应当确定项目设计文件中用来确定基线的假设、计算公式、原理以及其他信息都是可核实的,并应与其他可靠来源的信息相符。审定机构在判断可能的基线情形时,要充分考虑到项目所在地区的具体环境,采用可检验和可信的数据源(如当地专家的意见)来复核项目设计文件中提供的信息。

(e)审定机构应当参照所选方法学确定项目设计文件中的方程和参数是否被正确应用。如果方法学提供了选择的方程和参数的方法,则需确定项目参与方是否遵照执行。审定机构应当核实项目设计文件涉及的参数和数据的选择依据。如果数据和参数在拟议项目活动的计入期都保持不变,而且进行监测,审定机构应当审查所有数据来源和假设的合理性、计算的准确性、对拟议项目活动的适合性以及净温室气体汇清除量估算结果的保守性。如果在项目实施过程中对数据和参数进行了监测,且这些监测只有在对项目活动审定后才能获得,那么审定机构应该确认项目设计文件中对这些数据和参数的估计是合理的。

**4.3.2.7 确定管理活动的时间**

(1)审定要求。审定机构应当确定为了避免核查和碳库峰值产生系统性的重合,项目参与方是否必须参与计划管理活动包括收获循环和核查。

(2)审定方法。审定机构必须复查拟议项目活动的森林管理计划和监测计划,以避免核查和碳库峰值产生系统性重合。

**4.3.2.8 当地利益相关方的咨询过程**

(1)审定要求。审定机构应该确定项目参与方是否已经邀请了当地利益相关方对拟议项目活动进行评论。

(2)审定方法。审定机构应当通过文件评审以及必要时对当地利益相关方的访问确定:

- 所邀请的利益相关方是否对拟议项目活动做出了合理的

评论。

- 项目设计文件中提供的关于评论的摘要是否完整。
- 项目参与方是否对所获得的意见进行了解释,并在项目设计文件中进行了描述。

#### 4.3.2.9 监测计划

(1)审定要求。审定机构应当决定项目设计文件描述了监测计划,且该监测计划是基于拟议项目活动选择的、经批准的方法学制定的。

(2)审定方法。

(a)审定机构必须审定监测计划是否符合所选择的方法学的要求。

- 通过文件评审,识别所选择的方法学要求的参数。
- 确认监测计划是否包含了必要的参数,计划中所描述的监测方法是否符合所选定方法学。
- 由于间伐和主伐会导致碳储量降低,为使监测时间不与碳储量的峰值出现的时间重合,审定机构必须对间伐和主伐时间进行审定,以确保项目参与方所选定的核查期与碳库碳储量的峰值没有系统性的重复。

(b)审定机构还应当通过对记录的复审、有关人员的采访,项目计划,对拟议碳汇项目活动现场的考察来评价:

- 监测计划中描述的监测频率、监测时间等监测安排在项目设计中是否是可行的。
- 监测计划实施的方式,包括数据管理、质量保证和质量控制过程,是足以确保拟议项目活动产生的汇清除量可以被事后报告和核证。

### 4.3.2.10 环境和社会经济影响

(1)审定要求。审定机构必须确定项目参与方是否在项目设计文件中对项目活动所产生的社会经济和环境影响进行了分析,包括对项目区以及项目边界外的生物多样性和自然生态系统的影响。

(2)审定方法。审定机构应当通过文件评审和/或使用当地官方资源和专家的方式,确定项目参与方是否进行了社会经济和环境影响分析,包括对项目区以及项目边界外的生物多样性和自然生态系统的影响。这种分析应该包括当地社区、生物多样性、自然生态系统、土著居民、当地就业、粮食生产、文化与宗教等。

如果分析结果表明拟议的项目活动对环境无重大的负面影响,审定机构还需对项目所涉地方政府环境保护局所出具的项目环境影响评价报告(包括对生物多样性、水土流失的影响等内容)进行审定,以确认项目活动无重大负面的环境影响。如果项目使用引种,应该确认是否有证据证明该引种不是入侵种。

如果项目参与方认为项目有重大的负面影响,则审定机构必须通过文件评审的方式确定根据当地政府的相关规定进行了环境和社会经济影响评价,并在项目设计文件中对评价结果进行了描述。

## 4.4 审定意见

(1)审定机构应当对项目设计文件中预测的人为净温室气体汇清除量产生的可能性进行阐述。

(2)审定机构应当将审定结果告知审定项目的参与方。告知的内容应当包括:

(a)确认审定合格的信息以及递交审定报告给相关管理部门的日期。

(b)审定机构认为项目没有执行审定要求,不接受审定工作的理由。

(3)审定机构应当提供下列内容之一:

(a)在提交注册申请的审定报告中给出肯定的意见。

(b)在审定报告中给出否定的意见,并解释审定机构认为拟议项目活动不符合相关管理部门要求的理由。

(4)审定机构给出的审定意见应当包括:

(a)审定过程中采用的方法、过程以及标准的概要。

(b)项目审定过程中没有包括的内容和问题的描述。

(c)审定结论概要。

(d)对预期的汇清除量进行审定的陈述。

(e)对拟议的项目活动是否满足现有标准的意见。

## 4.5 审定报告

### 4.5.1 审定内容的报告要求

(1)项目描述。审定机构应当在审定报告中阐述对项目描述的准确性和完整性进行审定的过程以及对项目描述准确性和完整性审定意见。如果没有进行现场评审,需说明理由。

(2)林业碳汇项目活动边界。在审定报告中,审定机构必须陈述所评估的文件或者被访者的口述(如果有的话),并确定他们在中国的合法性。如果机构采用了抽样的方法,审定报告必须额外陈述评估了多少个项目点以及如何选择这些项目点的。

(3)碳库。审定机构必须在审定报告中声明:项目业主所选碳库是否符合所选定的经相关管理部门认可的方法学、方法学是否允许忽略某些碳库、项目业主是否选择了忽略这些碳库、忽略这些碳库对项目是否合理。

（4）土地合格性。审定机构应当在审定报告中详细叙述所评估的数据的来源以及现场审定过程的发现，以陈述对土地合格性进行审定的过程以及声明项目边界内的全部土地是否符合本指南中所指的土地合格性标准。

（5）非持久性和泄漏。审定报告必须描述项目业主用于减少非持久性和泄漏风险的方法。

（6）所选择的方法学的应用。审定机构应当在审定报告中清楚地表述根据所选方法学列出的每一个适用条件对项目设计文件中的相关信息进行评审的步骤。审定机构应当在审定报告中对拟议项目活动所选方法学的适用性做出明确的审定意见。

审定机构应当清楚地描述审定项目设计文件基线情景所采取的步骤，并明确对以下各方面给出审定意见：

——项目设计文件是否列出了项目参与方使用的假设和参数，包括他们引用的文献和数据源。

——项目参与方所使用的所有文件是否都和建立基线情景相关，并且在项目设计文件中被正确引用和解释。

——项目参与方确认基线情景所使用的假设和数据是否被证明是合适的，如有证据支持，可以被认为是合理的。

——项目设计文件中是否考虑并列出了相关的国家和林业行业政策及其存在环境。

——项目参与方是否正确应用了被相关管理部门（例如：国家林业局碳汇项目管理部门）认可的方法学来确定最合理的基线情景，且所确定的基线情景合理表现了没有拟议项目活动时会发生的情况。

——审定机构应当在审定中对审定和核查项目设计文件信息

所采用的步骤和所使用的信息来源进行详细的描述。

——是否正确地应用了方法学使所确定基线情景能够合理地代表没有拟议项目活动的基准净温室气体汇清除。

——是否正确地应用了方法学和相关的工具计算基准和净人为温室气体汇清除和泄漏。

——使用项目设计文件中提供的数据和参数值,是否可以重现所有基准温室气体汇清除的估计值。

(7)管理活动的时间(包括循环和核查)。审定报告必须阐述项目业主是如何避免核查和碳储量峰值产生系统性重合。

(8)监测计划。

——审定机构应当在审定报告中阐述其对监测计划是否符合方法学要求的意见。

——描述判断监测计划中的监测安排对于项目设计而言是否可行的步骤。

——审定报告应采用陈述项目参与方是否有能力实施监测计划的意见。

(9)当地利益相关群体的咨询过程。

审定报告应当:

——描述对当地利益相关群体的咨询过程是否为充分的评估过程。

——给出审定机构对当地利益相关群体咨询过程是否充分的意见。

(10)社会经济和环境影响(包括对生物多样性和自然生态系统的影响)。审定机构应当在审定报告中阐述项目参与方是否对社会经济和环境影响进行了分析,并在项目设计文件中对分析结果进行了描述。对于项目设计文件包括了可能产生负面影响结果的情况，还需阐述项目参与方是否在项目设计文件中描述了弥补负面影响的措施和监测安排。如果项目所在地的管理部门要求，还需阐述项目参与方是否提供了在符合当地法律法规的社会经济和环境影响评估报告。

### 4.5.2 审定报告

（1）审定机构应当在审定报告中包括最终的审定意见，报告应当：

（a）阐述拟议项目活动是否满足相关管理部门要求的审定结论。

（b）概述审定机构为得出审定结论和意见而开展的活动。

（c）包括审定机构与项目参与方对话的结果以及根据利益相关群体反馈的意见对项目设计文件所做的调整。报告应当反映对项目设计文件所做的讨论和修订。

（2）审定机构应当在审定报告中提供以下内容：

（a）审定过程和结论概要。

（b）得出所有结果和结论所采用的方法，特别是基线选择、排放因子、生物量转换因子以及监测的方法。

（c）审定机构在提交用于注册的审定报告之前所进行的全国范围内的利益相关群体的访谈信息，包括访谈日期、如何考虑反馈意见等。

（d）被访谈者名单以及评审文件清单。

（e）审定小组、技术专家以及国内技术评审专家的详细信息以及他们在审定过程中所发挥的作用，并指明现场评审工作的负责人。

（f）审定小组及审定过程的质量控制信息。

（g）审定小组成员的委托证书或简历。

# 5 核 查

## 5.1 核查的目的

(1)确保林业碳汇项目活动是按照其注册的项目设计文件运行的,项目所有的技术、监测和测量设备都已到位。

(2)确保监测报告和项目业主所提供的其他支持性文件是完整的、可验证的、与碳汇项目管理部门的要求相一致的。

(3)确保真实的监测体系和程序遵循在监测计划和所选的方法学中描述的监测体系和程序。

(4)用其监测方法学来评估存档和储存的数据。

## 5.2 核查的途径

(1)核查机构应当评价和核证项目活动的实施和所报告的人为净温室气体汇清除量的步骤遵循碳汇项目管理部门的相关规定或所认可的指南。

(2)核查机构应当对项目设计文件中涉及净温室气体汇清除量的定量和定性信息进行评估。

(3)核查机构应当评估和确认所执行的项目活动以及报告汇清除量的步骤与碳汇管理部门的要求或指定的指南相符合。这种评估应当包括对相关文件的复核以及现场核查。

(4)除了项目业主提供的监测文件,核查机构还应当复核。

——注册的项目设计文件,包括监测计划和相应的审定报告。

——以前的核证报告(如果有的话)。

——使用的监测方法学。

——任何其他与项目产生的人为净温室气体汇清除量相关的信息和参考(如 IPCC 的报告,国家森林资源清查报告和国家法规)。

## 5.3 核查的方法

核查机构应当用标准的审计方法来评估信息的质量,包括但不限于以下方法:

(1)文件的核查:

(a)核查项目参与方提供的数据和信息以检验其完整性。

(b)核查监测计划和监测方法学,尤其要注意监测频率、测量设备的质量,包括校核要求以及质量保证和控制程序。

(c)评估在项目设计文件中的管理数据、质量保证和控制体系的质量对碳汇量的产生和报告的影响。

(2)现场的核查:

(a)核查所实施的碳汇项目是按照已注册的项目设计文件实施和运行的。

(b)核查监测参数的产生、收集以及报告的信息流。

(c)对相关人员进行访问,以确定项目是按照项目设计文件中的监测计划实施运行和数据收集的。

(d)对监测报告中提供的信息和其他来源的数据进行复核。

(e)对监测和计量设备进行检查,根据项目设计文件和选择的方法学中的要求,观察监测活动。

(f)审核为确定温室气体汇清除量数据而做出的计算和假设。

(g)确认项目已经实施了质量保证和质量控制程序,以预防、识别并校正报告的监测参数中的任何错误或遗漏。

### 5.3.1 核查内容

(1) 核查项目业主所提供证据的质量。
(2) 核查项目的实施与已注册的项目设计文件的一致性。
(3) 核查监测计划与监测方法学的一致性。
(4) 核查监测活动与监测计划的一致性。
(5) 核查项目产生的人为净温室气体汇清除的数据和计算。

### 5.3.2 核查的步骤

(1) 项目实施与所注册的项目设计文件的一致性。

<u>核查要求</u>

核查机构应该确定实际开展的项目活动与已注册的项目设计文件的一致性以及相关所有问题，并确定：

(a) 项目活动是否已经根据项目设计文件中所描述的要求执行和实施。

(b) 所执行或实施的项目活动是否与项目标准的要求不同（包括偏离、提出新的建议或实际改变）。

<u>核查方法</u>

通过现场考察，核查机构应当确保已注册的项目设计文件中提到的拟议的林业碳汇项目活动全部物理特征都是到位的，且项目业主已按照注册过项目设计文件实施了拟议的项目活动，并考虑了与活动相关的指南。核查机构必须按照项目设计文件中描述核查拟议的碳汇项目活动的实施。如果没有进行现场核查，核查机构应该确定其决定的合理性。

(2)监测计划与监测方法学的一致性。

__核查要求__

核查机构应当核查项目活动实施的监测计划是否与已选用的方法学保持一致。

__核查方法__

核查机构应当确定项目是否已经按照已注册或批准的项目设计文件中的相关规定实施。

(3)监测活动与监测计划的一致性。

__核查要求__

核查机构应确定估算项目活动的汇清除量的参数是否已经按照已注册的项目设计文件中的规定进行了监测。

__核查方法__

核查机构应当确认：

(a)项目业主是否正确实施了监测计划。

(b)项目业主对监测计划中提出的所有参数都进行了监测，并根据具体情况及时进行了更新(如项目汇清除量和排放量参数、基准线汇清除和排放参数、泄漏参数、管理和运行系统，监测和报告的职责及权力与监测计划中规定的一致)。

(c)用于监测和计量的设备的精确度与碳汇项目管理部门制定的相关规定和指南相一致，并与监测计划中的控制和校核相一致(如根据经批准的频率持续记录监测结果；根据监测计划实施质量保证和质量控制程序)。

(d)监测结果是根据已批准的监测频率获得的。

(e)已经根据监测计划进行质量控制和质量保证程序。

(4)项目产生的人为净温室气体汇清除的数据和计算。

核查要求

核查机构应当根据所选择的方法学评估项目活动所产生的净温室气体汇清除量的数据和计算。

核查方法

核查机构应核查：

(a)项目业主是否已获得指定监测周期内的完整数据组。如果因为项目的水平而造成只能获得部分数据，或者非活动参数没有按照已注册的监测计划进行监测，核查机构应当确认最终的监测报告的结果是采用最保守方法获得的。

(b)监测报告中提供的信息是否与森林资源调查、购买记录和实验室分析等其他来源进行了相互核查。

(c)项目业主是否按照监测计划和方法学中描述的公式和方法，对基线情景的汇清除量、拟议碳汇项目活动的汇清除量和排放进行了计算。

(d)项目业主用于估算净温室气体汇清除的假设是否正确。

(e)项目业主是否正确地应用了合适的生物量转换因子、排放因子、IPCC缺省值以及其他参考数值。

## 5.4 核查报告

### 5.4.1 审定报告各项内容及要求

(1)项目的实施与已注册的项目设计文件的一致性。

对每一监测期，核查机构应当报告：

(a)项目的实施状态。核查机构应当清楚地描述每个造林地块的实施状态和项目开始实施的日期。对于分期执行的碳汇项目，核查机构应该在报告中指明，项目在每个核查阶段所取得的成果；如果执行期被延迟，核查机构应当描述理由以及预计执行的日期。

(b)项目活动的实际运行情况。

(c)监测报告中不同于已注册的项目设计文件中描述的、会导致目前监测周期内人为净温室气体汇清除量高于而且在未来的监测期也很可能会使汇清除量高于估计值的信息(数据和变量)。

(2)监测计划与监测方法学的一致性。

核查机构应当在核查报告中声明：监测计划和拟议的碳汇项目活动应用的已批准的方法学是一致的。

(3)监测活动与监测计划的一致性。

核查报告应当指明监测碳储量变化的活动是按照项目设计文件中拟定的监测计划实施的。

列出监测计划中要求监测的各种参数，阐述核查机构核查监测报告涉及参数的过程。

(4)项目产生的人为净温室气体汇清除数据和计算。

(a)核查报告应当指明特定监测期内是否能获得完整的数据。如果因为项目活动而造成只能获得部分数据，或者非活动参数没有按照已注册的监测计划进行监测，核查机构应当指明为了确保理论上的保守型原则而采取的活动。

(b)核查报告应当对复核报告中数据的过程进行描述。

(c)核查报告应证明，报告中计算基线情景、项目情景的碳汇量以及泄漏的公式和方法是合适的。

(d)核查报告应当对用于计算的假设、生物量扩展因子、排放因子、缺省值是否合理给出意见。

## 5.4.2 核查报告

为了获得最终的核查结论，核查机构应当对核查过程提出总的看法，并对核查过程中的发现进行清晰的说明和证明。此外，核查机构应当在核查报告中对所有支持性文件进行描述，如果有必要，需将其进行提供。

核查报告应当包括以下内容：

（1）核查过程和核查范围的概述。

（2）核查小组的详细信息（包括技术专家、内部评审员在核查过程中所起的作用以及现场核查负责人的详细信息）。

（3）文件评审和现场考察中的发现。

（4）核查机构的所有发现和结论，包括：

（a）项目活动是否按照已注册或已批准的项目设计文件的要求实施。

（b）监测计划是否符合监测方法学，监测活动是否按照监测计划执行。

（c）温室气体汇清除的评估。

（5）列出监测计划中的所有参数，陈述监测报告中数据被核查的过程。

（6）给出经核查后所获得的净温室气体汇清除量。

# 1 Scope of Application and Normative References

## 1.1 Scope of application

This guideline is applicable to organizations that are under contractual arrangements with project participants or coordinating /management entities to validate and /or verify any forestry-based carbon sequestration projects activities in China.

## 1.2 Normative references

The terms in the following documents will compose of the terms of this guidance when they are cited by this guidance. The revision of the following documents will not be fit for this guidance when they are put a date. The latest version, however, will be fit for this guidance when they are not put a date:

—IPCC: Good Practice Guidance for Land Use, Land-Use Change and Forestry;

—IPCC: Good Practice Guidance and Uncertainty Management in National Greenhouse Gas Inventories;

—2006 IPCC Guidelines for National Greenhouse Gas Inventories

—CDM Validation and Verification Manual

—Modalities and procedures for afforestation and reforestation project activities under the clean development mechanism in the first commitment period of the Kyoto Protocol.

# 2 Terms and definitions

The following terms and definitions apply in this guidance:

**Forest-based carbon sequestration project**

The projects that aim to increase or maintain forest carbon stock, including afforestation or reforestation project and sustainable management projects and deforestation.

**Forest:**

"Forest" is a minimum area of land of 0.05 ~ 1.0 hectares with tree crown cover (or equivalent stocking level) of more than 10 ~ 30 per cent with trees with the potential to reach a minimum height of 2 ~ 5 meters at maturity in site.

**Afforestation:**

"Afforestation" is the direct human-induced conversion of land that has not been forested to forested land through planting, seeding and/or the human-induced promotion of natural seed sources.

**Reforestation:**

"Reforestation" is the direct human-induced conversion of non-forested land to forested land through planting, seeding and/or the human-induced promotion of natural seed sources, on land that was forested but that has been converted to non-forested land.

**Revegetation:**

A direct human-induced activity to increase carbon stocks on sites through the establishment of vegetation that covers a minimum area of 0.05 hectares and does not meet the definitions of afforestation and reforestation contained above.

## Sustainable forest management

"Sustainable forest management" can be described as the attainment of balance-balance between society's increasing demands for forest products and benefits, and the preservation of forest health and diversity

## Deforestation

"Deforestation" is those practices or processes that result in the conversion of forested lands for non-forest uses.

## Project boundary

The "project boundary" geographically delineates the forestry-based carbon sequestration project activity under the control of the project participants. The project activity may contain more than one discrete area of land.

## Carbon pool

"Carbon pools" are those carbon pools: above-ground biomass, below-ground biomass, litter, dead wood and soil organic carbon.

## Baseline

"Baseline" is the sum of the changes in carbon stocks in the carbon pools within the project boundary that would have occurred in the absence of the forestry-based carbon project activity.

## Actual net greenhouse gas removals by sinks

"Actual net greenhouse gas removals by sinks" is the sum of the verifiable changes in carbon stocks in the carbon pools within the project boundary, minus the increase in emissions of the greenhouse gases measured in $CO_2$ equivalents by the sources that are increased as a result of the implementation of forestry-based carbon project activity, while avoiding double counting, within the project boundary, attributable to the forestry-based carbon project activity.

## Leakage

"Leakage" is the increase in greenhouse gas emissions by sources which occurs outside the boundary of the forestry-based carbon project activity, which is measurable and attributable to the forestry-based carbon project activity.

## Net anthropogenic greenhouse gas removals by sinks

"Net anthropogenic greenhouse gas removals by sinks" is the actual net-greenhouse gas removals by sinks minus the baseline net greenhouse gas removals by sinks minus leakage;

# 3 Principles for validation and verification

The following principles all shall be applied in performing validation and verification and all be used as guidance when documents related to validation and verification are prepared.

**Consistency**

Validating or verifying organization shall adopt same standards to validate and verify projects with similar characteristics (such as application of methodology, use of technology, time period and regional similarity), and treat opinions from experts for all projects. Feasible countermeasure complying with decides of government should be initiatively considered and set down if new decides are issued by government.

**Transparency**

Validating or verifyingorganizations shall disclose information to allow intended users to understand and to make decisions with reasonable confidence. Transparency relates to the degree to which information is seen to as being reported in an open, clear, factual, neutral and coherent manner based on documentary evidence.

**Independence**

Validating or verifying organizations shall remain independent of the activity being validated or verified and free from bias and any real or potential conflict of interest, and maintain objectivity throughout the validation and/or verification process to ensure that the findings and conclusions are based on objective evidence generated during the validation or verification and are not influenced by other interests or parties. .

### Impartiality

Validating or verifying organizations shall reflect truthfully and accurately validation or verification activities, findings, conclusions and reports. Report significant obstacles encountered during the validation or verification process, as well as unresolved, diverging opinions among validators or verifiers, the responsible part and the client.

### Confidentiality

Validating or verifying organizations shall ensure the confidential information obtained or created during validation or verification activities is safeguard.

# 4 Validation

## 4.1 Objective of validation

Validating organization shall conduct a thorough and independent assessment of proposed project activities against the applicable requirements to ensure approved projects meet the requirements of the department for forest-based carbon sequestration project management of SFA in China.

## 4.2 Validation approach

In carrying out its validation work, the validating organization shall:

(a) Assess the claims and assumptions made in the project design document. The evidence used in this assessment shall not be limited to that provided by the project participants.

(b) Ensure norelevant evidence that can change validation conclusion has been omitted. Determine whether the proposed project activity complies with the requirements of ruler set by the Department of carbon sequestration management of SFA (or CDM/VCS), and the applicability conditions of the approved methodology and guidance by Department of Carbon Sequestration Management, SFA.

## 4.3 Means of validation

(1) The organization shall assess the information provided by the project participants.

(2) In assessing information, the organization shall apply the means of vali-

dation specified throughout this standard and where appropriate standard auditing techniques, including, but not limited to:

(a) Document review, including:

(i) A review of data and information provided by the project participants.

(ii) Cross checks between information provided in the project participants and in the following information, if necessary, independent background investigations:

—National forest inventory in the host country (as applicable to the project area).

—Forest statistics.

—Growth models or yields table.

—Aerial photography, Satellite images, maps.

—GPS data.

—Historical land use/cover changes.

—Stratification and Sampling approach.

—IPCC GPG-LULUCF.

—Forest regulatory framework and land use policies of the host country (as applicable to the project area).

(b) Follow-up actions (e. g. on-site visit and telephone or email interviews), including:

(i) Interviews with relevant stakeholders in the host county/town/village, personnel with knowledge of the project design and implementation.

(ii) Cross checks between information provided by interviewed personnel (i. e. by checking sources or other interviews) to ensure that no relevant information has been omitted.

(c) Reference to available information relating to projects or technologies similar to the proposed project activity under validation.

(d) Review, based on the approved methodology being applied, of the appropriateness of formulate and accuracy of calculation.

### 4.3.1 Lists of validation

(1) The project design document

(a) Description of the proposed project activity.

(b) Boundary of the proposed project activity.

(c) Selection of carbon pools.

(d) Eligibility of the land that implement a proposed project.

(e) Approach for minimize risk of non-permanence and leakage.

(f) Application of the selected methodology, including:

—Baseline scenario identification and description

—Algorithms and/or formulae used to estimate amount of net anthropogenic GHG removals by sink over the chosen crediting period.

(g) Timing of management activity (including harvesting cycles and verification).

(h) Environmental and socio-economic impacts of the proposed project activity.

(2) Feasibility study report for the proposed project activity.

(3) Environmental impact assessment report for the proposed project.

(4) The proposed project activity implement plan.

### 4.3.2 Validation step

#### 4.3.2.1 Document review

(1) Description of the proposed project activity

<u>Validation requirement</u>

The validating organization shall confirm whether the description of the proposed project activity in the project design document is accurate, complete, and understanding of the proposed project activity.

Means of validation

The validating organization shall determine whether the following contents are described in the project design document, and whether the description is accurate, complete, and provides an understanding of the proposed project activity:

—General description of the proposed project activity.

—Information of project participants.

—Description of location of the proposed project activity.

—Description of the present environmental conditions of the area planned for the proposed project activity, including a brief description of climate, landform and topography, hydrology, soils, ecosystems and socio-economic state.

—Technical description of the proposed project activity.

—Description of legal title to the land, current land tenure and rights to carbon credit issued for the proposed project activity.

—Eligibility of the land.

—Approach proposed to address non-permanence.

—Carbon crediting period of the proposed project activity.

—The approach to assess carbon stock of the baseline scenario and the project scenario.

—Monitoring plan.

—Socio-economic and environmental impacts of the proposed project, including impacts on biodiversity and natural ecosystem.

—Stakeholder's comments.

(2) The project boundary

<u>Validation requirement</u>

The validating organization shall confirm whether the project design document contains a description of the project boundary that delineates discrete areas of land planned for the proposed project activities under the control of the project participants.

<u>Means of validation</u>

(a) The validating organization shall, through document review and/or interviews, determine the project design document contains a description including specific geographical positions (longitude, latitude) at each corner of planted parcel and geographical locations of counties/villages where the proposed project locates.

(b) The validating organization shall, through document review and/or interviews, determine whether the project participants for all areas of land planned for the proposed project activity:

—Have already established the control over activities; or

—Has the control over activities.

(c) The validating organization shall confirm that the control has included at minimum the exclusive right, defined in a way acceptable under the legal system of China, to perform the proposed activity with the aim of achieving net anthropogenic GHG removal by sinks. If the number of document to be reviewed and/or persons/entities to be interviewed is not less than 10, then the organization may apply a sampling approach.

(3) Selection of carbon pools

<u>Validation requirement</u>

The validating organization shall determine the carbon pools to be considered in the proposed project activity were selected in accordance with the requirements of the selected methodology.

Means of validation

The validating organization shall confirm that information has been provided to justify the exclusion of certain carbon pools if the methodology allows for such an option. In doing so, the organization shall confirm that all documents referred to in the project design document are correctly quoted and interpreted. If relevant, the organization shall cross-check in information provided in the project design document with other available information from public sources or local experts.

(4) Eligibility of land

Validation requirement

The validating organization shall confirm that the land within the planned project boundary is eligible for the proposed project in accordance with requirements of the department of the Department of Carbon Sequestration Management in SFA.

Means of validation

The validating organization shall confirm whether the land within the planned project boundary at the beginning of the proposed project is in accordance with requirements of the department of the Department of Carbon Sequestration Project Management in SFA based on a review of information and a site visit.

(5) Minimize potential risk of non-permanence and leakage

Validation requirement

The validating organization shall confirm that the project participants specified the approach selected to minimize potential risk of non-permanence and leakage.

Means of validation

(a) The validating organization shall review the project design document to ensure an approach to minimize risk of non-permanence and leakage is selected according to the relevant provisions of the Department of Carbon Sink Project Management in SFA, or IPCC GPG for LULUCF.

(b) The validating organization shall determine whether the approach that minimizes risk of carbon reverse result from harvest, fire, disease and deforestation is description (E. g. buffer approach).

(6) Application of the selected methodology

Validation requirement

(a) The validating organization shall determine whether the baseline and monitoring methodologies selected by the project participant are the valid versions of those approved by Department of Carbon Sink Project Management in SFA, or the approach proposed by IPCC GPG for LULUCF.

(b) The validating organization shall determine whether the baseline identified for the proposed project activity is the scenario that reasonably represents the net anthropogenic GHG removals by sink that would occur in the absence of the proposed project activity.

(c) The validating organization shall determine whether the steps taken and the equations and parameters applied in the project design document to calculate project net GHG removals by sink, baseline net GHG removals by sink, leakage comply with the requirements of the selected methodology including applicable tools.

Means of validation

(a) The validating organization shall determine whether the methodology is correctly quoted and applied by comparing it with the actual text of the applicable version of the methodology.

(b) The validating organization shall determine whether the project activity meets each of the applicability conditions of the approved methodology or any tool or other methodology component referred to therein. This shall be done by validating the documentation referred to in the project design document. If the validating organization, based on local and forestry sectoral knowledge, is aware that comparable information is available from credible sources other than that used in the project design document, then the organization shall cross-check the project design document against other sources to confirm that the project activity meets the applicability conditions of the methodology.

(c) The validating organization shall determine whether the proposed project activity result in GHG emission in addition to the provision of the selected methodology.

(d) If the methodology requires several alternative scenarios to be considered in the identification of the most plausible baseline scenario, the organization shall, based on financial expertise and local and forestry sectoral knowledge, determine whether all scenarios that are considered by the project participants and any scenarios that are supplementary to those required by the methodology, are realistic and credible in the context of the proposed project activity and that no alternative scenario has been excluded. The organization shall determine whether the most plausible baseline scenario identified is reasonable by validating the assumptions, calculations and rationales used in the project design document. It shall determine whether documents and sources referred to in the project design document are correctly quoted and interpreted. The organization shall cross-check the information provided in the project design document with other verifiable and credible sources, such as local expert opinion, if available.

(e) Algorithms and/or formulae used to estimate amount of net anthropogenic GHG removals by sink where the methodology allows for selection between options for equations or parameters, the validating organization shall determine whether adequate justification has been provided (based on the choice of the baseline scenario, context of the proposed project activity and other evidence provided) and that the correct equations and parameters have been used, in accordance with the methodology selected including applicable tool(s). The validating organization shall verify the justification given in the in the project design document for the choice of data and parameters used in the equations. If data and parameters will not be monitored throughout the crediting period of the proposed project activity but have already been determined and will remain fixed throughout the crediting period, the organization shall determine whether all data sources and assumptions are appropriate and calculations are correct as applicable to the proposed project activity, and will result in an accurate or otherwise conservative estimate of the emission reductions. If data and parameters will be monitored or estimated on implementation and hence become available only after validation of the project activity, the organization shall determine whether the estimates provided in the project design document for these data and parameters are reasonable.

(7) Timing of management activity

Validation requirement

The validating organization shall determine whether the project design document describes the planned management activities, including harvesting cycles, and verifications such that a systematic coincidence of verification and peaks in carbon stocks would be avoided

Means of validation

The validating organization shall review the forest management plan and the monitoring plan for the proposed project activity to confirm that a systematic coincidence of verification and peaks in carbon stocks is avoided.

(8) Local stakeholder consultation

Validation requirement

The organization shall determine whether the project participants have completed a local stakeholder consultation process and that due steps were taken to engage stakeholders and solicit comments for the proposed project activity.

Means of validation

The organization shall, by means of document review and interviews with local stakeholders as appropriate, determine whether:

(a) Comments have been invited from local stakeholders that are relevant for the proposed project activity.

(b) The summary of the comments received as provided in the project design document is complete.

(c) The project participants have taken due account of all comments received and have described this process in the project design document is complete.

(9) Monitoring plan

Validation requirement

The validating organization shall determine whether the description of the monitoring plan included in the project design document is based on the selected monitoring methodology including applicable tool(s).

Means of validation

(a) The validating organization shall assess compliance of the monitoring plan with the selected approved methodology.

(i) Identify the list of parameters required by the selected approved methodology including applicable tool(s) by means of document review.

(ii) Confirm that the description of the monitoring plan contains all necessary parameters, that they are described and that the means of monitoring described in the plan complies with the requirements of the methodology including applicable tool(s).

(iii) Validate thinning time and final cutting time described in project design document to ensure systematic coincidence of verification and peaks in carbon stocks would be avoided.

(b) The validating organization shall assess the implementation of the plan by means of review of the documented procedures, interviews with relevant personnel, project plans and any physical inspection of the proposed project activity site, assess whether:

(i) The monitoring arrangements (E. g. monitoring frequency) described in the monitoring plan are feasible within the project design.

(ii) The means of implementation of the monitoring plan, including the data management and quality assurance and quality control procedures, are sufficient to ensure that the emission reductions achieved by/resulting from the proposed project activity can be reported ex post and verified.

(10) Environmental and socio-economic impacts of the proposed project activity

Validation requirement

The validating organization shall validate the documentation received from the project participants on its analysis of the socio-economic and environmental impacts, including impacts on biodiversity and natural ecosystems, and impacts in and outside the project boundary of the proposed project activity.

Means of validation

The validating organization shall determine whether the major environmental and social-economic impacts of the proposed project activity outside the proposed project boundary have been analyzed in the project design document by means of a document review and/or using local official sources and expertise. This analysis should include, where applicable, information on, *inter alia* local communities, biodiversity, natural ecosystems, indigenous people, local employment, food production, cultural and religious.

If any negative impact is considered no significant by the project participants, an environmental impact assessment report (including biodiversity, soil impact…) have been undertaken by local environment protect bureau should be reviewed that to affirm negative impact of the project activity is no significant.

If non-native species are used by the project, identify whether these non-native species are not invasive species.

If the above-mentioned analysis leads to the conclusion that a negative impact that may be considered significant by the project participants has been detected, then the organization shall determine whether a socio-economic impact assessment and/or a environmental impact assessment has been undertaken in accordance with relevant regulations, and the outcome of such impact assessment is summarized in the project design document.

## 4.4 Validation Opinion

(1) The validating organization shall include a statement of the likelihood of the project activity achieving the anticipated amount of net anthropogenic GHG removals by sink stated in the project design document.

(2) The validating organization shall inform the project participants of the validation outcome. Notification to the project participants shall include:

(a) A confirmation of validation and date of submission of the validation report to the relevant management department; or

(b) An explanation of reasons for non-acceptance if the project activity, as documented, is determined not to implement the requirements for validation.

(3) The validating organization shall provide either:

(a) A positive validation opinion in its validation report that is submitted as a request for registration; or

(b) A negative validation opinion in its validation report explaining the reason for its opinion if the organization determines that the proposed project activity does not implement the applicable requirements.

(4) The organization shall include the following in its opinion:

(a) A summary of the validation methodology and process used and the validation criteria applied.

(b) A description of project components or issues not covered by the validation process.

(c) A summary of the validation conclusions.

(d) A statement on the validation of the expected amount of net anthropogenic GHG removals by sink.

(e) A statement as to whether the proposed project activity meets the stated criteria.

## 4.5 Validation Report

### 4.5.1 Report requirements for validation list

(1) Description of the proposed project activity

The validating organization shall:

(a) Describe the process undertaken to validate the accuracy and completeness of the project description.

(b) Provide an opinion on the accuracy and completeness of the project description.

(c) Provide a justification if it has not conducted a site visit.

(2) Project boundary

The validating organization shall describe the documentation assessed and/or oral statements delivered by persons interviewed (if any) and determine their acceptability under the legal system of China. If the organization has applied a sampling approach, it shall also describe how many sites have been assessed and how these sites were selected.

(3) Selection of carbon pools

If the methodology allows for the option to exclude certain pools and this option is selected by project participants, the validating organization shall provide a statement as to whether the selection of carbon pools complies with the selected methodology, and whether the exclusion is justified.

(4) Eligibility of the land that implement a proposed project

The validating organization shall describe how the validation of the eligibility of the land has been performed, by detailing the data sources assessed and by describing its observations during the site visit. The validating organization shall provide a statement as to whether the entire land within the project boundary is eligible for a proposed project activity.

(5) Approach for minimize risk of non-permanence and leakage

The validating organization shall confirm whether the approach selected by the project participants to address non-permanence has been specified in the project design document.

(6) Application of the selected methodology

(a) For each applicability condition listed in the approved methodology selected, the validating organization shall describe the steps taken to assess the relevant information contained in the project design document against these criteria. The organization shall provide a validation opinion regarding the applicability of the selected methodology to the proposed project activity.

(b) The validating organization shall describe the steps taken to assess the requirements and provide an opinion as to whether:

(i) All the assumptions and data used by the project participants are listed in the project design document, including their references and sources.

(ii) All documentation used is relevant for establishing the baseline scenario and correctly quoted and interpreted in the project design document.

(iii) Assumptions and data used in the identification of the baseline scenario are justified appropriately, supported by evidence and can be deemed reasonable.

(iv) Relevant national and/or forestry sectoral policies and circumstances are considered and listed in the project design document.

(v) The approved baseline methodology has been correctly applied to identify the most plausible baseline scenario and the identified baseline scenario reasonably represents what would occur in the absence of the proposed project activity.

(vi) The baseline methodology and corresponding tool(s) have been applied correctly to calculate the net anthropogenic GHG removals by sink, GHG removals by sink and leakage.

(vii) All estimates of the baseline emissions can be replicated using the data and parameter values provided in the project design document.

(7) Timing of management activity

The organization shall describe how the project participants have ensured that a systematic coincidence of verification and peaks in carbon stocks would be avoided

(8) Monitoring plan

The validating organization shall:

(a) State its opinion on the compliance of the described monitoring plan with the requirements of the methodology including applicable tool(s).

(b) Describe the steps undertaken to assess whether the monitoring arrangements described in the monitoring plan are feasible within the project design.

(c) State its opinion on the project participant's ability to implement the described monitoring plan.

(9) Local stakeholder consultation

The validating organization shall:

(a) Describe the steps taken to assess the adequacy of the local stakeholder consultation.

(b) Provide an opinion on the adequacy of the local stakeholder consultation.

(10) Environmental and socio-economic impacts of the proposed project activity.

The validating organization shall confirm whether the project participants have undertaken an analysis of the socio-economic and environmental impacts and, if required by the local government, a socio-economic impact assessment and/or an environmental impact assessment in accordance with relevant local regulations.

The validating organization shall also note whether the outcome of such impact assessment has been summarized in the project design document and whether a description of the planned monitoring and remedial measures to address the negative impacts has been included in the project design document.

## 4.5.2 Validation report

(1) The organization shall include the final validation opinion in the valida-

tion report. In its validation report, the organization shall:

(a) State its conclusions regarding the proposed project activity's conformity with applicable requirements of relevant management department.

(b) Give an overview of the validation activities carried out in order to arrive at the final validation conclusions and opinion.

(c) Include the results of the dialogue between the organization and the project participants, as well as any adjustments made to the project design following stakeholder consultation. It shall reflect discussions on and revisions to project documentation.

(2) In its validation report, the organization shall provide the following:

(a) A summary of the validation process and its conclusions.

(b) All its applied approaches, "findings and conclusions, especially on baseline selection, emission factors, BEF and monitoring".

(c) Information on the nationally stakeholder consultation carried out by the organization prior to submitting the project for validation, including dates and how comments received have been taken into consideration by the organization.

(d) A list of interviewees and documents reviewed.

(e) Details of the validation team, technical experts, internal technical reviewers involved, together with their roles in the validation activity and details of who conducted the on-site visit.

(f) Information on quality control within the team and in the validation process.

(g) Appointment certificates or curricula vitae of the organization's validation team members, technical experts and internal technical reviewers for the project activity.

# 5 Verification

## 5.1 Objective of verification

The verifying organization that is approved by relevant management department for carbon sink project shall conduct a through, independent assessment of the registered project activities to ensure:

(1) The proposed project activity was fulfilled in accordance with the registered project design document.

(2) Monitoring report and support document provided by the project participants are complete, validate and meet the requirement of relevant management department.

(3) Actual monitoring system and procedure comply with requirements of the selected methodology.

(4) Dates were recorded and archived according to the selected monitoring methodology.

## 5.2 Verification approach

(1) In carrying out its verification work, the verifying organization shall determine whether the project activity complies with the requirements and guidance provided by the department of carbon sink project management.

(2) The verifying organization shall assess both quantitative and qualitative information on emission reductions provided in the project design documentation.

(3) The verifying organization shall assess and determine whether the imple-

mentation and operation of the project activity, and the steps taken to report the net anthropogenic GHG removals by sink comply with the relevant guidance provided by the department of carbon sink project management This assessment shall involve a review of relevant documentation as well as an on-site visit(s).

(4) In addition to the monitoring documentation the verifying organization shall review:

(a) The registered project design documentation and the monitoring plan, including any approved revised monitoring plan and/or changes from the registered project design documentation, and the corresponding validation opinion.

(b) The validation report.

(c) Previous verification reports, if any.

(d) The applied monitoring methodology.

(e) Any other information and references relevant to the project activity's net anthropogenic GHG removals by sink (e. g. IPCC reports, data onnational forest inventory or laboratory analysis and national regulations);

## 5.3 Means of verification

The verifying organization shall apply standard auditing techniques to assess the quality of the information, including but not limited to:

(a) Desk review, involving:

(i) A review of the data and information presented to verify their completeness.

(ii) A review of the monitoring plan and monitoring methodology, including applicable tools, paying particular attention to the frequency of measurements, the quality of metering equipment including calibration requirements, and the quality assurance and quality control procedures.

(iii) An evaluation of data management and the quality assurance and quality control system in the context of their influence on the generation and reporting of emission reductions.

(b) On-site assessment, involving:

(i) An assessment of the implementation and operation of the registered project activity as per the registered the project design document or any approved revised the project design document.

(ii) A review of information flows for generating, aggregating and reporting the monitoring parameters.

(iii) Interviews with relevant personnel to determine whether the operational and data collection procedures are implemented in accordance with the monitoring plan in the project design document.

(iv) A cross check between information provided in the monitoring report and data from other source such as plant logbooks, inventories, purchase records or similar data sources.

(v) A check of the monitoring equipment including calibration performance and observations of monitoring practices against the requirements of the project design document and the selected methodology and corresponding tool(s), where applicable.

(vi) A review of calculations and assumptions made in determining the GHG data and emission reductions.

(vii) An identification of quality control and quality assurance procedures in place to prevent or identify and correct any errors or omissions in the reported monitoring parameters.

### 5.3.1 Verification list

(1) Compliance of the project implementation with the registered project de-

sign document

(2) Compliance of the monitoring plan with the monitoring methodology.

(3) Compliance of the monitoring activities with the monitoring plan.

(4) Assessment of data and calculation of net anthropogenic GHG removal by sink.

## 5.3.2 Verification of compliance

(1) Compliance of the project implementation with the registered project design document

Verification requirement

(a) The verifying organization shall identify any concerns related to the conformity of the actual project activity and its operation with the registered project design document and determine whether:

—The implementation and operation of the project activity has been conducted in accordance with the description contained in the registered project design document; or

(b) Any deviation or the proposed or actual changes in the implementation or operation of the project activity comply with the requirements of the Project Standard.

Means of verification

The verifying organization shall, by means of an on-site visit, assess that all physical features of the project activity in the registered project design document are in place and that the project participants have operated the project activity as per the registered project design document or any approved revised project design document. If an on-site visit is not conducted, the verify organization shall justify the rationale of the decision.

(2) Compliance of the monitoring plan with the monitoring methodology

Verification requirement

The verifying organization shall determine whether the monitoring plan of the project activity is in accordance with the applied methodology including applicable tool(s).

Means of verification

The verify organization shall determine whether the project implementation is in accordance with the provisions of the registered project design document and/or an approved revised project design document.

(3) Compliance of the monitoring activities with the monitoring plan

Verification requirement

The verify organization shall determine whether the monitoring of parameters related to the GHG emissions reductions in the project activity has been implemented in accordance with the monitoring plan contained in the registered project design document.

Means of verification

The verify organization shall determine whether:

(a) The monitoring plan has been properly implemented and followed by the project participants.

(b) All parameters stated in the monitoring plan and relevant Board decisions37 have been monitored and updated as applicable, including:

(i) Project removal or emission parameters.

(ii) Baseline removal or emission parameters.

(iii) Leakage parameters.

(iv) Management and operational system: the responsibilities and authorities for monitoring and reporting are in accordance with the responsibilities and authorities stated in the monitoring plan.

(c) The equipment used for monitoring is in accordance with requirements of the department project management and is controlled and calibrated in accordance with the monitoring plan, the applied methodology, local/national standards, or as per the manufacturers specification.

(d) Monitoring results are consistently recorded as per approved frequency.

(e) Quality assurance and quality control procedures have been applied in accordance with the monitoring plan.

(4) Assessment of data and calculation of net anthropogenic GHG removal by sink.

Verification requirement

The verifying organization shall assess the data and calculations of GHG net anthropogenic GHG removal by sink achieved by/resulting from the project activity by the application of the selected approved methodology.

Means of verification

The verifying organization shall determine whether:

(a) A complete set of data for the specified monitoring period is available. If only partial data are available because activity levels or non-activity parameters have not been monitored in accordance with the registered monitoring plan, the verification shall ensure that the most conservation option have been taken.

(b) Information provided in the monitoring report has been cross-checked with other sources such as inventories, purchase records, laboratory analysis.

(c) Calculations of baseline removals or emissions, and project activity removals or emissions and leakage, as appropriate, have been carried out in accordance with the formulae and methods described in the monitoring plan and the applied methodology document.

(d) Any assumptions used in removals calculations have been justified.

(e) Appropriate biomass exchange factor, emission factors, IPCC default values and other reference values have been correctly applied.

## 5.4 Verification report

### 5.4.1 Report requirements for verification list

(1) Compliance of the project implementation with the registered project design document

For each monitoring period, the verify organization shall report:

(a) The implementation status of the project. For project activities that consist of more than one site, the verify organization shall describe the status of implementation and starting date of operation for each site. For project activities with phased implementation, the verify organization shall state the progress of the proposed project activity achieved in each phase under verification. If the phased implementation is delayed, the verify organization shall describe the reasons and present the expected implementation dates.

(b) The actual operation of the project activity.

(c) Information (data and variables) provided in the monitoring report that is different from that stated in the registered project design document, and has caused an increase in estimates of the removal increase in the current monitoring period or is highly likely to increase the estimates of emission reductions in the future monitoring periods.

(2) Compliance of the monitoring plan with the monitoring methodology

The verify organization shall provide a statement whether the monitoring plan is in accordance with the approved methodology applied by the registered project activity.

(3) Compliance of the monitoring activities with the monitoring plan

The verify organization shall state whether monitoring has been carried out in accordance with the monitoring plan contained in the registered project design document.

The verify organization shall list each parameter required by the monitoring plan and state how it verified the information flow (from data generation, aggregation, to recording, calculation and reporting) for these parameters including the values in the monitoring reports.

(4) Assessment of data and calculation of net anthropogenic GHG removal by sink

The verification report shall contain:

(a) An indication of whether data were not available because activity levels or non-activity parameters were not monitored in accordance with the registered monitoring plan as well as any actions taken by the DOE to ensure that the most conservative assumption theoretically possible has been made.

(b) A description of how the DOE cross-checked reported data.

(c) A confirmation that appropriate methods and formulae for calculating baseline removal or emission, project removal or emissions and leakage have been followed.

(d) An opinion as to whether assumptions, biomass exchange factors, emission factors and default values that were applied in the calculations have been justified.

## 5.4.2 Verification report

The verification report shall give an overview of the verification process used by the verifying organization in order to arrive at its verification conclusions. All verification findings shall be identified and justified. The verifying organization shall describe all documentation supporting verification and shall make it available on request.

The verification report shall report the following:

(a) A summary of the verification process and the scope of verification.

(b) Details of the verification team, technical experts, and internal reviewers involved, together with their roles in the verification activity and details of who conducted the on-site visit.

(c) Findings of the desk review and site visit.

(d) All of the verifying organization's findings and conclusions as to whether:

(i) The project activity has been implemented and operated in accordance with the registered project design document or any approved project design document.

(ii) The monitoring plan complies with the monitoring methodology and the actual monitoring complies with the monitoring plan.

(iii) The data and calculation of GHG removal by sink have been assessed to correctly support the amount of GHG removal by sink being claimed.

(e) A list of each parameter specified by the monitoring plan and a statement on how the values in the monitoring report have been verified.

(f) A conclusion on the verified amount of removal by sink achieved.

# References

1. IPCC. Good practice guidance for land use, land-use change and forestry (Task 1) [M]. Japan: IGES. 2003.
2. IPCC. Good practice guidance and uncertainty management in national greenhouse gas inventories. Workbook [M]. Japan: IGES. 2000.
3. IPCC. 2006 IPCC Guidelines for nation greenhouse gas inventories: Workbook [M]. Japan: IGES. 2006.
4. UNFCCC. 2012. Guidelines for completing the project design document form for afforestation and reforestation CDM project activities (version 01.0). FCCC/EB/2012/66/Add10. http://cdm.unfccc.int/EB/archives/meetings_10.html#66
5. UNFCCC. 2012. Clean development mechanism validation and verification standard (version 02.0) FCCC/EB/2012/65/Add10. http://cdm.unfccc.int/EB/archives/meetings_10.html#66
6. UNFCCC. 2003. Modalities and procedures for afforestation and reforestation project activities under the clean development mechanism in the first commitment period of the KyotoProtocol. UNFCCC/CP/2003/6/Add.2, 2003. http://unfccc.int/resource/docs/cop9/06a02.pdf